BUSINESS DECISIONS WITH LEAN AND SIX SIGMA

HOW TO IMPROVE YOURSELF AND YOUR BUSINESS

Izaac Mack

Copyright 2022. All Rights Reserved.

This document provides exact and reliable information regarding the topic and issues covered. The publication is sold with the idea that the publisher is not required to render accounting, officially permitted, or otherwise qualified services. If advice is necessary, legal or professional, a practiced individual in the profession should be ordered. From a Declaration of Principles which was accepted and approved equally by a Committee of the American Bar Association and a Committee of Publishers and Associations. In no way is it legal to reproduce, duplicate, or transmit any part of this document in either electronic means or printed format. Recording of this publication is strictly prohibited, and any storage of this document is not allowed unless with written permission from the publisher. All rights reserved. The information provided herein is stated to be truthful and consistent. Any liability, in terms of inattention or otherwise, by any usage or abuse of any policies, processes, or Instructions contained within is the solitary and utter responsibility of the recipient reader. Under no circumstances will any legal obligation or blame be held against the publisher for reparation, damages, or monetary loss due to the information herein, either directly or indirectly. Respective authors own all copyrights not held by the publisher. The information herein is offered for informational purposes solely and is universal as such. The presentation of the data is without a contract or any guarantee assurance.

TABLE OF CONTENT

INTRODUCTION 5

CHAPTER 1 - UNDERSTANDING THE GAME 8

CHAPTER 2 - THE VITAL STRATEGY OF LEAN SIX SIGMA INVESTMENT 22

CHAPTER 3 - BE FASTER TO IMPROVE YOURSELF 27

CHAPTER 4 - THE AIM OF SIX SIGMA 30

CHAPTER 5 - WHY DOES LEAN, AND SIX SIGMA NEED EACH OTHER? 42

CHAPTER 6 - BUSINESS DECISION 53

CHAPTER 7 - CULTURE OF CONTINUOUS IMPROVEMENT 65

INTRODUCTION

Lean 6 sigma is the formation of two established methodologies responsible for training excellent professionals who act as change agents. For a better understanding, six sigma tilt and continuous improvement. Lean is an efficient manufacturing/process concept from the toyota production system in the mid-twentieth century and aims to dispose of waste. This depends on the principle of determining the value from the customer's point of view and always improving the way the cost is presented, which eliminates any use of resources that may be ineffective or not add value. Lean focuses on maintaining value with minimal work on providing the ideal product to the customer through the process of creating a zero-waste value. This is done by allowing each worker to realize his full potential and thus make the best possible contribution. The goal of empowerment is based on the idea of showing respect to people, but it extends beyond the final customer and can include workers, suppliers, and society. Lean strives to improve value proposition, reduce process losses, maximize human potential, and allow employees to improve their work continuously. Lean assists employees to develop professionally and personally, allowing them to be proud of their work, and an important characteristic of this methodology is value creation, where the value is created - known as gemba. To understand lin's participation in lean six sigma, we need to understand lin's fundamental principles presented in "the machine that changed the world (1991)," by James p. Womack, Daniel T. Jones, and Daniel Ross. The authors studied many manufacturing systems and also wrote the book based on their notes at Toyota. Based on these studies, the following principles were identified: determine the value from the viewpoint of the final consumer, according to the product family. identify all steps in the value stream for each product family (what are the steps to obtain the final product), removing steps that do not produce value where possible. follow the quick steps to create the value so that the product easily moves to the customer.when the flow is released, allow customers to take the process to the final customer when the flow is defined, and the missing steps are deleted, the new flow is used. Repeat the process and continue until the perfect condition is reached where there is no more waste. Womack and jones recommend that supervisors and CEOS involved in the lean app be aware of three things: Objective: what are the customer problems that the company will solve? Procedure: how the organization evaluates each value stream to ensure that all measures are valid, capable, available, sufficient, and flexible and that all steps are related to the flow and production extraction. When settling? People: how does the company ensure that everything that has been improved does not disappear in the flow of the product? How do we ensure that people participate in the continuous improvement of these flows? With the speed of lean, people can easily understand the need for lean six sigma and its basic truth: quality improves speed, and speed improves quality. But i heard time and time again a question: how does lean six sigma apply to a service organization?

Chapter 1 - UNDERSTANDING THE GAME

Lean six sigma return on investment in services "the initial lack of six sigma priority in non-industrial areas was a wrong selection that cost Motorola industry at over $ 4 billion in four years." the operation service now accounts for over 80% of u.s. gdp and is growing rapidly all around the world. Even for manufacturers, it is common for only 20% of product prices to depend on direct industrial employment - the remaining 80% are the costs involved in the output or related to design function and support. (product development, human resource, engineering, purchasing, and finance, etc.). In addition, for services, labor costs without value-added ("no value-added") are higher in percent and absolute dollars than in production. Revenue growth potential associated with speed and quality of service often hides savings opportunities. For example, a company that does not add value to its customers typically accounts for 50% of the total cost of the service. This represents an enormous potential for "white-collar workers" to achieve significant improvements in quality, cost, and speed, which can give companies a great logical benefit over their competitors. When graham Richard, a businessman, and entrepreneur, was elected mayor of fort Wayne, Indiana, he had a simple vision: "I want fort Wayne to be a safe city. I want him to get good jobs. I want to have great service and attract new business. He knew the city could not continue the "bureaucracy as usual" if it implemented this vision, but was there an alternative operating in the government? Examples of known organizations that need to implement lean six sigma in their business operations and services: just like others, bank colleagues, bank one materialized several times during the 1990s, which means mergers and acquisitions mean that heROIC efforts were needed daily in order to fulfill the main mission. In a competitive sector such as finance, this situation cannot stay for a while despite the long journey ahead of them. At stanford hospital and clinics (shc), the future was clear: patient sizes decreased as shc continued to lose contracts due to higher costs. Doctors and management realized that if they didn't do something quickly, they would still lose existing patients and be unable to attract new patients. One thing is the desire to provide quality care to patients, but pragmatism works under this motto: "there is no margin, there is no task." Although these companies were discovered to come from various departments, they are a symbol of important opportunities for putting lean six sigma in practice. Their objectives and aim are not completely similar, and so every individual need also vary from logistical assistance to medical support medical for manufacturing, but all are at the forefront of new development. They discovered that the best way to accomplish their goals was to integrate the principles and methods of lean and six sigma to improve the operation of service. Bank one began to use these methods and principles as a guiding principle in its national business operations, themed focus 2.0. It started in February of the year 2002 and started with a group of properly selected projects with strategic importance. Through its efforts, the neo group discovered the chance to generate millions of revenues annually through process improvements and savings a lot of dollars, which sum up to thousands at the end of the year by avoiding costs and reducing losses in other operations. Lockheed martin has set a vision: "we want lean 6s operations." they can identify a big list of operations, from design purchases, which now take a small fraction of the time and costs they are used to. In fact, more than 1,000 completed tasks in recent years within the service areas without including other sectors. Their debt has dropped, revenues are good, they will exceed the goal of cutting costs, and they discovered a record number of arrears. They were able to deliver their latest missiles (with all the resources necessary to the customer) at a lesser cost up to half of what it used to be and for the previous years, one-third cycle due to the increase in the use of lean six sigma, and don't use cheaper materials or sharp corners! They were able to win a joint strike contract, valued at more than $ 100 billion. Mike joyce, vice president of lockheed martin, said: "numerous parameters contribute to this type of outcome, but one of the main contributors is lm21 (21st century lockheed martin), which is our initiative for the organizational efficiency based on lean six sigma in the operational process." The mayor of fort

Wayne graham Richard permitted the launch of several projects throughout the city, according to the principles and methods of lean and six sigma. Many municipal services have noted improvements in certain aspects of their services to citizens (complaints avenue, faster response, clearer contact times to requests for information), a clear reduction in better use of the city's resources, or costs. For example, the modification of building permits reduced the response time from around two to around two months and eliminated this type of problem, which prevents many organizations from the urge to do a business transaction with the city. Development in the collection of garbage has reduced costs by approximately $200,000 per annum for the subordinate contractor in the process of providing a much better service. Every organization has recognized several basic facts: Complexity reduction helps to improve quality and speed. This course only occurs if you apply six sigma and lean together. Improving quality can really make it faster Increasing speed can really improve quality Lean six sigma services are methodologies for businesses used for improvement while increasing shareholder value and achieves a rapid rate of development in terms of investment in capital, quality, cost, speed of operation, customer satisfaction. The integration of lean and six sigma optimization methods is necessary for the following reasons: six sigma alone cannot significantly improve the speed of the process or reduce the capital invested the statistical control does not give room for lean operation Ironically, lean and six sigma are often seen as competing initiatives - lean fans note that six sigma shows little concern to anything speed related agenda and flow. On the other hand, six sigma supporters note that lean did not approach concepts based on differences and customer needs. Either side is correct. However, these divergences are used basically to pick one ahead of second, instead of supporting the logical conclusion we need to integrate six sigma and lean. How is the six sigma lean and complement of one another?

Six sigma
Pay much attention to the importance of identifying opportunities and eliminate errors, as defined by customers Realizes that the difference undermines

our energy to provide quality services with reliability Requires decision driven by data and integrates a complete group of quality tools into a robust framework to effectively solve the problem Provide highly effective and normative cultural infrastructure for an efficient outcome When used the right way, it promises to offer an enhanced operating profit of $ 500,000 or more for every black belt per annum (a strong number that many companies regularly reach)

Lean
It's main focus is to optimize the velocity of the process Identify the right tools for process flow analyses and delay times for every activity taking place in the process Focuses on separating "non-value added value" work from "value-added" it use tools to remove the causes and costs of activities with no added value Provides a way to measure and remove complexity cost Both lean and six sigma approaches reinforce and interact with each other so that gains in the ROIC% are faster if six sigma and lean are applied together. (others may ask whether the return on investment is a valid measure for service companies, the answer is definitely yes:

numerous service restaurants, healthcare, health care, and hotels require a large capital. Other service companies -financial services, government, software development, etc. - the most important costs are salaries/benefits, so the capital invested is really "people's cost.") In summary, what distinguishes lean six sigma from its components is the core recognition that you cannot deliver only speed or just quality. Six sigma and lean have their roots in the 1980s (and later) at a time when there was a high demand for quality and speed in industrialization. Lean has emerged as a vehicle improvement industry. Six sigma has developed to a quality initiative targeted at eliminating imperfections, reducing process contrasts in the semiconductor industry. It is therefore not surprising that the first six sigma and lean service applications appear in the function of service support for companies who deal with manufacturing - caterpillar, ge capital, lockheed martin, finance, itt, etc. These organizations were already specialized in six lean skill and six sigma: data collection, value flow mapping, variation analysis, experiment design, and composition reduction. It is impossible for foreigners to know to what extent the reported profits at these organizations are a result of improvements in manufacturing compared to improvement in service operations Relationship between speed and quality About 30 to 50% of the costs in a service organization are due to costs related to slow speed or tracking to meet customer needs. The development of the value calculations provides a way to demonstrate calculation that only a fast, interactive process is capable of achieving the highest levels of quality, and only a high-quality process that can maintain a high speed. Six sigma + lean only = lower cost Most people in the working environments have come across six sigma and a targeted improvement method to achieve an extremely high level of quality that helped them to generate millions of ads to the bare minimum of companies such as allied signal, ge, lockheed martin, and iit. On the other hand, only a few people that are not used to manufacturing are familiar with lean applications. What happens people hear the word "lean" and automatically think of "industrialization" because that is how it is always used. Lean also creates process speed reduced cycle time with efficiency cost, minimum time, and capital invested in each process. Just as it is in the six sigma, lean has become a manufacturing company, but lean seems even more comfortable in the factory than what is known of the six sigma. Terms 5s, like sale of workstations, kanban, batch processing, wip, configuration, and cloud systems, have no inherent meaning for those whose task is talking to customers, writing on a computer, with coordinating services. However, these concepts have strong applications in services. However, lean six sigma has one basic principle, which is unnecessary complexity, and it leads to huge time, costs, and enormous waste. A practical analysis was carried out to make a group of surgeons work together to develop a common platform for surgical operations. Of course, they were initially skeptical. But if you tap on it to find out more than the others at close range, surgeons found that six different trays had little impact on the quality of patient care. During certain interactions, they were able to agree with one another on a standard surgical operating platform. The experiment was later added to lean six sigma tactics. The result? As previously mentioned, the annual costs of materials have decreased. Here is another example from fort Wayne. The department of transport has an annual budget of $ 2 million. Once a large spending proposal has been adopted, these funds are "held" according to their terms and are no longer available for other purposes. It sounds like the right process, doesn't it? But what if a tight budget arrives? This meant that no money was needed, but the department could not use it for other tasks. Or say that the excess labor costs the city has to fight for money it might not have. In the beginning, the problem team found a huge gap between the actual price and the offer price and the actual price of most jobs - about 23% less than the offer at 22% more. The initial stage was to create a bid target of ± 10% at your real cost. Is this the right option? Nobody knew it at first! The fact was that the city didn't have a goal before.) By reversing the process determining the workflow and annulling complex activities and using the data to identify the problems, the team was able to reduce the variance in the amounts of the offers, which largely frees up $ 200,000 per year, which are already "blocked." Translate manufacturing to service Using the lean point of view, the lack of funds in an environment that delivers services, such as the production capacity not available in the factory, has the same consequences. Showed that device malfunctions with regard to changing requirements cause numerous workflows (wip) to be advanced and thus delay in completing this work. In fort Wayne, the lack of funds had a similar impact: tasks with very different costs were made to remain in effect (delay in completion) until funds were available. These are typical examples of the benefits that have a high possibility of accomplishment by using six sigma and lean tools together in application services. Three main reasons reason has been identified as the major factors why you should apply the service functionality to the lean six sigma application: service operations are often slow and expensive. Sluggish processes will be subject to low quality, which leads to higher costs and reduces customer satisfaction and thus, sales. The outcome of slow processes: the larger part of the costs for service applications are waste without added value. Service processes are slow due to too much work in progress (wip), usually due to useless complexity in the provision of the product/service. It doesn't matter whether wip is a pending report on the table, email in a mailbox, or customer orders stored in the database. When a lot of work is done, the company can dedicate over 90% of the waiting time, which does not help its customers and actually causes or adds significant waste (worthless costs). Who does the lean method apply to? The lean tools and methods are useful for whoever; tracks information to perform a task ("lack of information" in service corresponds to a lack of equipment in production) it must go through several decision cycles continues to interrupt when finding a way to complete the task acceleration (for reports, procurement, materials, etc.) i do not know what they do not know works in batches (adding multiple items that require the same type of work before starting the corresponding tasks) searching for lost work in "empty space" around the company storage In any slow process, 80% of the delay is due to less than 20% of the activities. Finding and improving the speed of 20% of the process steps is enough to reduce cycle time by 80% and reach more than 99% at the time of delivery. In summary, those who work in services discover that most stages of their operations do not add value to the service delivered in the sight of their customer. Identify and add waste without value-added implementing lean six sigma, and the outcome will be followed as follows the next evening. It is not a cost reduction, but a change in the process.

Chapter 2 - THE VITAL STRATEGY OF LEAN SIX SIGMA INVESTMENT

In manufacturing, remarkable investment in the equipment may be needed to improve labor productivity. On the other hand, according to warren buffett, service operations are mainly determined by intellectual capital: "the best type of investment is an investment in which a high return results from a very small increase in invested capital" (berkshire hathaway) 1984). With the lean six sigma application, the ROIC equation counter can be increased without increasing the financial investment. In the lockheed martin procurement center, for example, the main investment to reduce acquisition costs by 50% was reimbursed for five months. Significant savings were made at stanford hospital and clinics by bringing together a group of non-capital surgeons. If buffett likes this type of investment, he will represent his shareholders. The concept of linking lean six sigma's efforts to critical shareholder value is rarely discussed. If the link is not created, your company can make a profit. However, it is important that your investment in lean six sigma helps you achieve your strategic goals. To illustrate the principle of increasing shareholder value, Tom Copland, a well-known agency that assessed companies whose references were identified during the Mckinsey consultant, compiled the actual value of the stock market for several hundred companies, including data on the book market value (vertical axis) measures the stock market premium stock market pays on the company's equity (book value). Some companies negotiate their book value once, others six times. Economic profit (horizontal y-axis), which denotes the "difference" between the return on investment (ROIC%) and the percentage of capital expenditure (COC%) c these assets are invested in treasury bills and risk exposure to stocks. Income growth y axis, which is another important factor for shareholder value when the economic benefits are positive, Economic profit (EP) = 0%, ROIC = company's capital expenditure, and experimental stock market data shows that the company is trading at approximately book value. If ROIC% is above 5% of the cost of capital, the company negotiates 4-5 times the carrying amount. Companies that can grow more than 10% per year at 10% are likely to do ten times more than their book value. Therefore, maximizing return on investment is very valuable. This is a vital point to make the six sigma projects prioritized according to the judgment of the P&L manager, depending on the company's ability to invest its energy into the ROIC ratio.

Sales growth pays off for shareholder value

In addition to increasing return on investment, reducing costs, and capital investment, lean six sigma plays an important role in growing sales. This only applies to organizations or processes that generate more profit than the cost of investment. Here is the primary argument of the finance: from the speech of a billionaire, warren buffet: "the value of a business is determined by inflows and outflows - discounted at a reasonable interest rate" (Berkshire Hathaway annual report, 1992). This quote may seem strange, but it is not. What buffett means is that the dollar he earned next year is less than the dollar he earned this year due to inflation. For example, if inflation is 5%, the "current" value of $ 1.00 next year is only $ 0.95. Reduced value = $ 1 / (1+ .05) = $ 0.95 The dollar earned in two years is $1/(1 + 0.05) 2 = $ 0.90 and so on. The idea of the discounted amount is vital as it has impact on sales growth, which should be seen as the "present value of the increased cash flow" (yes, you can expect a sales Increase of $ 100,000 following year the true value for your business, however, should only be presented as an inflation rate of 5% at $ 95,000, this principle corresponds to the present economic value gains. When comparing potential lean six sigma projects, you need to get feedback on those projects in 3-5 years and present them at (current) discounts using optimal growth rates. This is the only way to compare the performance of different projects, and lean six sigma projects have a significant impact on shareholder value. Is overall sales growth good? You can see that wasteful growth does not create value. They were going straight on the horizontal axis (= growth without profit). The curve hardly increases. That is the value of the increment increases. Lean six sigma services achieve results quickly. The types of outcomes that can be mitigated to support strategic goals. How to make their customers happy, they want to do more business, create shareholder value, and motivate their employees. What explains the quick results? Lean six sigma uses the principles of speed and immediate action into the six sigma improvement process. This improves the speed and results of development projects. Lean six sigma also focuses on variations of six sigma to reduce the impact of latency. Finally, lean six sigma specifically addresses the unexpected costs of bidder complexity.

Chapter 3 - BE FASTER TO IMPROVE YOURSELF

Why do you need lean and six sigma?

Hospitalization is a very important aspect of health care, which jeopardizes service providers: how to cut costs but with good service care. In the heart unit of stanford clinic and hospital, a process flows analysis revealed capacity limits in the "descending unit required by post-operative patients that have no requirement intensive care any longer. The limited capacity of this unit has ensured that patients stay in the costly unit for care longer than necessary. Instead of increasing the number of nurses or allocating spaces to expand the environment, a group checked the logs used in the removal unit, checked the determinant factor for readiness to be discharged, and resident factors that helped to stay longer to detect if anything can solve or prevent these problems in the hospital (e.g., changing the guidelines for the use of certain medications). Several changes were made, and they all lead to an increase in the capacity of the reduction unit without any significant investment peed. Low cost, quality. These global goals exist if there is competition in the business. In the past, a bank change initiative has always been associated with negative exchanges. A group of people dedicated their time to examine the cost, time, taught, and quality, "i can't improve all three." all business unit managers made independent decisions in a silo and never had to make team decisions. There is a cfo who cares about the costs, the manager assigned for change who wants to get good outcomes, and the special support to the task, for which time of completion is of importance. From my observation of lean six sigma, i can conclude that there is a pattern by which everyone can understand easily. People are beginning to know that they can get all three. It is no longer an exchange conversation. "but he talks about how you can benefit from them by combining all. This appears to be the first time that you can involve everyone. The players sit around the player in a good mind for speed, cost, and quality." various disciplines have developed to accomplish a similar aim. It all comes that the point that lean six sigma can only apply to all three at the same time because lean is combined with the main pivot on six sigma and process speed and with the main focus on the quality of the process. If you consider lean and six sigma competing practices, the central feature you cannot reach top speed without improving quality or achieving premium but not improving speed. You can only maximize return on return (ROIC) if you implement the two in your process.

Chapter 4 - THE AIM OF SIX SIGMA

Many approaches have been developed for the improvement of quality. Why should it be implementation leaders like Starwood hotels, quest diagnostics? Ge capital, admit six sigma as their favorite tool of operation? It must admit that no other tool demand such a brilliant list of supporters. One of six sigma's easiest and most powerful concepts gives an answer to this. The process results are a result of what's going on in the process. You can find this term in the simple equation "y is an x function" that links the (y) output to the variables of the processor input(xs): $y = f(x_1, x_2, x_3, ...)$ The above equation also applies at organizations each product (y), which could be growth, profit, return on investment, depends on input variables (xs), which could be time, quality, gravity, and undervalue - adding additional costs, etc. To improve our outcome, we need to research and concentrate on x critics who influence this result. This equation has a more meaning you will learn as you become more inclined to studying lean six sigma. It's not just "y is." some xs' job - but our job is to find the x value that drives y. Would you like your profit to increase? What entries do you need to influence to do? Would you like to improve the quality of any of the services you offer? What are the main service entries? Most affect the quality? If leaders could implement this equation, it will help to shape their behavior. There will be a stop to their 10% increase improvement request. Instead, they will believe in lean six sigma's efforts to help employees review and improve the processes that lead to this result. A review performed by Lou Giuliano at the p and I center, he first requests a black belt project presentation. The CEOS concentrate on xs, which defines ys contribution to the improvement cycle and identifies that continuous development is a fundamental tool in business.

The basic elements of six sigma recipe

Six sigma is an important tool used by business owners in the development of their business and making vital business decisions. If proper attention is given to It, It will guide the organization in making a valuable business decision to overcome competitors. The necessary tools are; Commitment from the ceo and management. The company has a group of shareholders whose ROIC is the main objective, a group of materials to use the core creation opportunities. The cost, speed, and quality of lean six sigma are ROIC drivers. For this reason, the ceo should be a leader in supporting the ideas, and the reason for the failure of the p&l director to "climb" should not be view as a possibility. The ceo must communicate regularly and demonstrate the commitment to the process of change, and all members of the management team must pass through formal leadership training. It is necessary to allocate appropriate resources (= employees and reduce time) for high priority projects. A six sigma promise is that a black belt full-time can generate $ 500,000 a year from the biggest operating profit that has a way of been traced to the end. Two notable parameters of this equation are the nature and number of resources used to define projects companies reach out to. The types of achievements generally engaged 1% of the best option leaders at work - as a full-time option. 3% of staff received belt training green. They also developed rigorously. The process - usually guided by the company sample - to define, define and select projects on the basis of criteria, which could be maximizing the return on investment (ROIC) and/or the possible impact on critical customer quality problems. All interested or involved in six sigma must receive a level of training. All managers and executives must be informed of six sigma. Training extent depends on the individual or group involved to be involved in the selection, direction, management, or implementation of the improvement. The change must be discarded. Contrast reduction is a woven concept into the texture and flesh of the six sigma foundation. The discrepancy in meeting customers' critical quality requirements (CTQ) is the main driver of the optimization process. Attack and contrast elimination is achieved through (Define measure analyze improvement control dmaic) troubleshooting tools and support tools that need the organization to make data-driven decisions.

What lean can do: low cost and speed

While six sigma is closely related to quality flaws and contrast elimination, lean is associated with waste elimination, efficiency, and speed. Lean's goal is to accelerate the speed of any process, reducing waste in all its forms. The overall advantage of lean felt in its ability to discover opportunities to reduce costs and deadlines never seen before. When applying lean tools and concepts, you will discover that the process steps that i previously thought were necessary are not necessary and that their costs and delays have been applied to removable tools after lean. You will begin to detect the difference between targeted models and practices and the ones that add interest to your customers. Here is an example: any company that thought of building in fort wayne was soon warned: dealing with the city was difficult

The origin of the terms used in six sigma

Six sigma terms derive from the interaction of the diversity of a process and customer needs attached to that process. The highest concentration of values is around the mean (mean) and is uniformly separated. The distance between the midline and the reflection point (where the curve begins to flatten) is known as sigma, the standard deviation. Sixty-eight percent of the data is within the standard deviation above or below average, 95% over 2 and 99% over

3. (therefore, a range of -3 to + 3 represents 99% of the data.) Six sigma numbers represent how the distributor actual output content is compared to the range of acceptable values (customer specifications). The defect is any value that is not within the customer's specifications.

Lean primer

Each specialty is identified by its own tool. Lean also have its own tools. There are some terms that you will find necessary to understand and appreciate the benefits of lean.

Delivery times and operating speed

The time limit is the time it takes to get your product or service delivered once the order has been processed. Understanding the reasons for a timeout is much easier than another process, using a simple equation little's law. Lead time = amount of work in the process/average completion rate This formula will be used to detect the duration of time needed to complete any component of the job (time limit) by simply calculating the amount of work that is yet to be completed (work in progress) with the amount of tasks we can do daily, weekly, etc. (average rate of return). Little's law is much more useful than it appears. Many of us have no idea of the average delivery time or delivery times, not to mention the difference. The idea of tracking the stage of the order is frightening, and it becomes more challenging when the process takes days or weeks to complete. But lead time makes it more convenient to calculate an even longer period of time. The lean principle is somehow tricky yet incredibly powerful lessons that allow us to make quick wins with lean. The following are the rules: Most operations are "scanty The main goal is to reduce the control over the wip (if you cannot control the wip, you cannot control the delivery times) Each process must operate in a draw, not pressure, to eliminate the difference in delivery times 80% delay is caused by just 20% of the activities Unseen work cannot be helped: there is a need for a visual, data-driven management Maybe you won't be surprised when you know that in "meager" service operations, most work, at the minimum, 50%, and sometimes has no added value. This point is easily illustrated by using colors or other techniques to separate the added value to work visually without adding value to the process map.

Take a look at little law:

Delivery time = amount of work in progress / average completion rate

This equation is not just a theoretical construct. It has much usefulness in operations. Firstly, it demonstrates that two methods for controlling delivery times are to reduce the work in the process (WIP) or increase the average of the completion rate in any process that doesn't deal directly with customers, this is where wip places orders, calls, emails, or reports, it is not people: controlling wip is much easier than improving the completion rate. Also, it is possible to reduce the timeout, increase any process, reducing the amount of wip, even if nothing is done to improve the completion of the rate. This conclusion explains how people can make these rapid gains by applying lean principles. Where possible, they should simply specify the amount of work they are allowed in the process of a particular time. Why should we focus on wip first? It only costs intellectual capital to reduce wip. It requires investment in finance capital or paychecks to increase the average rate of return, which damages both the return on investment and the value of the shareholders. Slim tools can reduce labor during the process and eliminate waste, thus improving yield. Little's law provides the mathematical foundations that allow us to apply lean operations to all operations. Take a quick view at the workspace around you, do you see a list of emails that just come in that might take days to reach? Do you have a mailbox full of floods? Does the answering machine reject the most recent messages? Are there people who are interested in your work delivery? All of these elements represent the work in progress (wip), which is the job that co-worker, a customer, colleague, customer, or any other person requires from you. For a newly developed lean thinker, it is important to limit the work in progress to have the opportunity to improve the time of action and reduce waste. Because you understand work in progress has a similitude of a car on a highway: when you add more cars, you don't get to increase the speed of the busy highway! But how do we do it? Obviously, it is not possible to end wip in customer operations when it is really about people who are waiting for their service or are trying to buy the product in these cases, and there are other ways to improve or maintain a deadline. For any company that isn't a real customer in your presence, the idea of limiting wip is in little's law. During lean service process, a particular step that precedes the real process, which is a step in which inputs are collected (job requests, requests, calls, etc.). Another person then handles the freeing of these "substances." Let's take a look at an example. Independent distributors of a company must obtain information on suggestions from the marketing department to cite construction work. The workers were dissatisfied with two or three weeks, and the marketing process required the development of the requested information. The shift required to make these customers happy took three days. A group of organized data in a matter of weeks showing that staff in the marketing department could have the chance to process up to 20 quotes in a day, distributors wanted a reliable three-day shift. The data showed that due to the difference in the process, the marketing team should achieve closer target days to satisfy this customer request. What amount does wip allow in this process? They switched to little law and put 20 in the percentage of completion with 2.4 days in the time limit to reach a maximum of 48 offers in progress at any given time of operation. Lead time = 2.4 days = the amount of work in the process = 48 quotes/average completion rate = 20 quotes / day In order to manage this process, they designed a visual box showing the number of offers processed. The maximum number allowed in work in progress was 48 requests, so, only if the number drops to 47, the employee cannot provide additional quotes. The given requests are unofficial in the process before they are let loose from the container of the raw materials. The only trigger to start an action in the process is to have an element that exits the process: this is the dragging system. The guaranteed level of service of approximately two and a half days does not begin until the request enters the process. In other words, a system of withdrawals in service environments means making informed decisions about the time of work released in the process. The way you make these decisions is critical, as it offers another chance to try to focus on "value." for example, in this case, the question was the request that will be reissued in the process after completing another request. "first, first," you didn't stop it here since some of the orders were very likely, with the possibility of a higher dollar; he was not likely to accept others, or offer difficult offers or lesser orders. The points are multiplied by each quota in each quotation opportunity. The next to be issued in the process will be the highest score, even if other bidding opportunities have been waiting for a long time. (therefore, a new order representing nine has been issued before the old order marked With this process, an equal number of marketers were able to reserve 70% of additional revenue, with 80% of the total profit. Obviously, the alternative to this company was to recruit more marketing team, at a higher cost, to make it possible to increase their capacity.

Chapter 5 - WHY DOES LEAN, AND SIX SIGMA NEED EACH OTHER?

Why does lean need six sigma

Since lean's strength is facing delivery times and costs that don't add value, there are many critical issues that won't be encountered in lean's half-year books. Six sigma offers powerful solutions to these problems, which explains why lean needs six sigma: Lean does not explicitly describe the culture and infrastructure needed to achieve and maintain results most of lean's resources change in the infrastructure needed to successfully implement lean's initiatives and achieve and accelerate sustain lean. This is true that many companies that implemented lean were developed to develop an infrastructure similar to that of six sigma, but they did it in an ad hoc way, rather than using the indicative framework included in six sigma. Sometimes the companies that implement lean are impossible to spread within the organization and maintain the results because they do not have the well-defined six sigma cultural infrastructure to generate the commitment of senior management, formalized training, guarantee allocated resources, etc. Therefore, the meager progress depends on the individual initiative. I have seen many successful apps for lean undo when hiring a new manager

Customer's critical quality needs are not in the foreground and center

By asking us to determine what "added value" is in the process, lean incorporates some elements of customer focus but is introspective in his approach. The person who creates the value flow map decides whether the activity has added value. In contrast, six sigma imposes innumerable positions in optimization methods, in which the ratings of customers and suppliers should be included. The customer uses critical-to-quality as the primary metric and requires a method for acquiring vocs in the dmaic identification phase. In simple terms, the customer is not the center of lean's interface but is always present in six sigma's works.

Does not recognize the effect of contrast

Lean does not have the tools to reduce contrast and put a process under statistical control. Six sigma claims that eliminating the difference is the key to the difference in demand generates a greater impact on how long your job should wait "on the waiting list" as the process approaches capacity (as shown by the sharp increase in the right curve). The greater the contrast, the greater the effect. Let's say bob takes an average of 16 minutes to do a particular job. However, due to the variety of tasks, 68% (standard deviation) of operation times can vary by about 8 minutes on both sides, in which case the contrast is 8/16 = 50%. Suppose also. The workload presented to bob has a similar difference. If bob is loaded with up to 90% capacity, the function supported behind him will wait an average of 60 minutes, which explains about half the waiting time. Having a particularly bad problem can easily free up that waiting time for more than 100 minutes. The change has little effect on low-energy operations (left side of the graph). But most service organizations operate within or near them total capacity - occurs when the change has a significant impact on the time that the job (or customer) has to wait for "inline." customer support processes often experience many differences in demand because we have no control over when customers will contact us. The lesson? The greater the variation in the input, the greater the power. If there is low variability or we can control demand in one way or another (which is more likely in internal operations), we can work with greater capacity without running the risk of excessive delays. When i presented this analysis to lockheed martin, manny zulweta (deputy head of the mac-mar supply center at lockheed martin) said, "this is what we saw!"

Why does six sigma need lean?

Because of the gap in lean methodology that can benefit from six sigma, the general message is: as many companies show, six sigma can make a lot of money. But there are obstacles. If you do not have flexible parts, you can choose any tool. If you don't focus on accelerating and reducing wip, your profits will end. The process is slow and complicated, and its cost is very high. More specifically, here are five reasons to use six sigma marine: Identification of waste. Process allocation refers to six sigma tools but does not describe the data collection processes (setup time, processing time per unit, transfer, etc.) Required to determine the steps that contribute significantly to different values. Process. Value. Add labor/cost to services or products. Lean provides a powerful tool for value flowcharts. Emphasizes multifunctional storage, waste, and delay. Six sigma rarely discusses classification operations as value-added, and eliminating value-added activities is not an important six sigma principle. Instead, the six sigma protocol initially claims to eliminate the difference, and if that is not possible, the process can be redesigned with six sigma design (dfss). If the cycle efficiency is less than 10%, the process must always be redesigned (to eliminate value-added activities). Improve process speed or cycle time: Improvements and corrections in cycle time are called six sigma results. However, six sigma's six experts (and their books) do not make a practical or theoretical connection between quality and speed, and the lab will be able to manage the wip size (how to manage lead time parameters). It does not collect system plugs, but the variations are limited. The main factor in the cycle is the continuous process (due to little law). If wip is not initially limited to extreme values, reducing cycle time is only an illusion. Speed limit tool. Excellent product management tools, time value analysis, 5s, etc. Are rarely found. Included in six sigma toolkit. These acceleration tools are very powerful and have been developed and enhanced with decades of experience. While it is true that these tools require translation to adapt to the service environment, ignoring them carries the risk of limiting operational performance. Quick action method (dmaic kaizen process): The lean methodology includes a quick improvement method called kaizen. Kaizen is a small, intensive project that invites a group of people with relevant knowledge for 4-5 days, applying systematic improvements to the targeted process or activity. The energy generated by these events is legendary and creates a high level of creativity through the pressure to achieve concrete results quickly. Kaizen plays a true role in a service environment but requires some changes. However, incorporating work-based optimization techniques into the arsenal provides the right accelerator for dmaic projects. Lynn's work accelerates results. The quality of six sigma can be handled very quickly if lean ignores the non-value added steps. Six sigma lab has created a chart that examines the performance of products in stock and analyzes the cumulative effect of errors on yield. For example, consider a billing system that includes 20 companies. All companies are operating at a rate of 4 (99.379%). The total return on compressed transfer pricing (.99379) is 20 = 88%, which is not common in service operations. This low yield raises credit issues and increases the need for manual purchase and restructuring procedures. Learn tools, and it's realistic to quickly remove activities (within a few weeks at most) that don't add value - at least half (= 10) steps. Therefore, instead of running a 20-step process, billing now only takes ten steps. Even without making other quality improvements, it seems clear that a 10-step process has far less chance of error than a 20-step process return will increase the transfer fee to (.99379) 10 = 94%. Higher returns will provide a significant return on investment in improvement and, more importantly, the speed of the process will double: this not only allows you to deliver the results to customers faster but also doubles the rate of comments when using quality. Tools, which make them effective twice. Combining lean and six sigma, the steps cannot be removed, but they improve the quality levels of each remaining activity by, for example, which raises the scrolling performance to (.99976) 10 = 99, 8%. Sigma lawyers challenge. The question arises whether it is better to optimize the process first (without removing the steps from the process without adding value) using six sigma or removing the steps without the added value through lean first and then improving the process through six sigma. Some proponents of six sigma suggest that lean methods such as trawling systems should only be implemented after the process has been controlled and improved. This view is easily contested: "how can this be detrimental to anything that uses lean and a pull system application so that you can control speed and reduce cycle times while applying six sigma?" the answer is that you work simultaneously with the lean six sigma toolkit and corporate culture to be the best. Projects should be chosen evaluated using their operation on the ROIC, not on solving a problem that requires multiple lean or six sigma tools. Six sigma mix to improve service The truth Is that lean six sigma is a powerful tool for implementing the CEO's strategy and a tactical tool for P&L managers to achieve their annual and quarterly goals. If executives are not involved in lean six sigma, the company is likely to outperform companies sponsored by lean six sigma. Integration topics for lean and six sigma provide five "laws" that guide our improvement efforts. Here are the first four (which we calculate from "0" because the first list is essential for all others): market law: the customer with critical quality determines the quality and represents the main priority for improvement, followed by return on investment and net present value. We call this law zero because it is the basis of everything else. the law of flexibility: the speed of any process is proportional to the flexibility of the process 2: law of concentration: 20% of the activities in the process cause an 80% delay. law of speed: the speed of any process inversely proportional to the amount of work in progress (or the number of things in progress). little's law states: turn, the number of objects in progress has increased due to long installation times, rebooting, the effect of changes in demand and supply, time, and complexity in supplying the product is one last rule for six sigma;

law of complexity and costs: the complexity of offering services or products generally adds additional costs with no added value and wip compared to low quality (low sigma) or low speed (not lean) processing problems. Remember that the costs that Stanford university achieved were realized at the same time that there was a significant decrease in mortality. The services through the eyes of your customers: a customer-centric organization was not a perfect market value of the shares, of the receivables from customers or of the fixed assets that exposed the distinctive candy yield rates of see, but rather a mixture of intangible assets, in particular a positive and widespread consumers reputation based about many interesting benefits. Their experiences with the product and the employees. This reputation generates a privilege that allows the buyer to evaluate the product, not the cost of production, as the main determinant of the selling price. " Hathaway annual report from 1983 and i, completely unaware of the importance of the economic franchise, analyzed only $ 7 million in see on the net as tangible rights (and 4.2 million in profits) and claimed that $ 25 million (not $ 30 million) was the highest we could get (and understand). Fortunately, the deal was not hampered by our senseless $ 25 million insistence. Over the past twenty years, see's has distributed $ 410 million to Berkshire hathaway to increase only undistributed earnings million dollars! " Hathaway 1991 annual report. Buffett knows, in his view, that the value derives from "fixed rates of return higher than the capitalization cost." It is a view supported by the stock market data. To offer quality, therefore, we must learn to observe the customer behavior of our customers and learn to understand what we do and not in the way they judge us. Discover a vision of what it is when awareness of the customers permeates all aspects of operations of the business. Integrate customer information into strategic decision making, use of voc data in the design of products/services, interweave voc data in process improvements, work model/set of skills based on customer needs of volatile organic compounds.

Chapter 6 - BUSINESS DECISION

Business decisions a broader level, understanding your customers also the place of the meeting they create must be a pivotal part of your organization's market point and organization goals and decisions. There different types of customer information are useful for this: your current products/services meet (or not) customer needs what are the needs of existing and currently not satisfying customers (market opportunity) what unnecessary customers offer (analysis of the service line/ product) to compare your offers with the competition are the overall performance levels (measurement)? The first lesson to learn from all of the above is to make sure you speak your language. "we sit down with our customers (measure everything) and discuss their parameters," by Manny Zoluita, vice president of the mac-mar shopping center in Lockheed martin. The unknown decision is of the voc team are two basic ways to collect information about vocs: get out and get them (proactive methods) or allow them to reach you (interactive methods). Methods medium of getting information to you on a customer's initiative. It covers customer complaints, requests, sales, calls compliments, technical support, access to web pages, e-mails or cards sent by customers, pos survey cards that fill them, contractual negotiations, references, and so on. Advanced methods for collecting, tracking, and using this information are absolutely essential to retain existing customers. This is to give an opinion on the current product. Because customers are more likely to contact you when they encounter problems or questions, interactive methods are more effective at finding weaknesses in products/services than benefits. It can also be distorted, as it represents more customer segments than other customer segments. These methods mean leading customer contact. It includes surveys/surveys, focus groups, interviews, site visits or visits, in-store contacts, etc. If you want to control the time and content of your contacts, you can use proactive methods for more purposes than interactive methods, such as product/service design, process improvement, performance monitoring brand market analysis. In recent years, it has become increasingly common to include one or two customer representatives on a problem/process resolution team. That data is "costs." therefore, if you are investing time and money in collecting volatile organic compound data, be sure to use that investment wisely, obtaining accurate and trusted customer information, ease of use for employees, and easy access to them. These metrics are still important; how do they prioritize them? That way, be sure to measure something important to them. Customer cash metrics form the basis for integrating our initiatives with our customers. "it was done well; the results we achieved contribute directly to achieving the goals of our customers. Our endless goal is always to exceed their expectations". An example of how this information is used is shown in the "bubble." there are five assessed flows in this company (in this case identical to the brands) and has simply started to analyze its strategic position by collecting customer voice data ("how much of our offers are similar to the competition"). This research can focus on any combination of price, jobs, related support services, etc. Y-axis shows whether the industry as a whole is producing economic profit for each product or service. The size of the circles reflects that the brand's revenue; the position on the chart is a combination of competitive position and market profitability. Brand a is in the best position: it creates returns (it's a big circle), are in a profitable market (indicated at the top of the chart) and competitive (on the right side of the chart). The e mark is in the opposite position: it is small and is in the bottom left corner of the chart (which means it has low revenue and is uncompetitive in a non-profit sector). Series of products/services in which the sector as a whole makes a negative economic profit is 'unattractive sector.' this does not mean that no company is doing well, but the media is not. (for example, from the beginning, the aviation industry had a negative economic profit. But companies with a competitive advantage, like southwest airlines, made positive economic profits.). Service/product evaluation advanced use of information from voc depends on a large model of the market; here, the focus is much more restricted, on customer feedback on specific projects, features, and functions of a specific product or service, and other there are two situations that this information usually requires are: Whether existing services/products meet critical quality needs (ctq) Collect voc data to create design requirements for new or redesigned products/services The two situations use aims to understand the best method of satisfying customers and what could not satisfy customers if implemented in the business process. Does your business strategy meet the desire of the customer? Do not buy a product or service because of the general market trends. Interact with functions or features. That is why you are always there trying to check the functions and features or whatever product or service you are offering meets the customer demand. By implementing the voc decision design, one can prevent some conditions from occurring by weaving the customer's voice into your product/service design decisions. The generally acceptable six sigma methodology adapted is the QFD (quality function deployment), this is a methodology that is used for changing the customer desire to a particular feature of service or product plan. In order to accomplish this, there are two primary stages, identifying the voc i.e., knowing what's critical for customers and using the QFD to change what the customer initially needs to the requirement that is functional and later to the design requirements. The product or service process design must begin with the customer's voice – whatever the customer actually want from your service. The distribution of quality functions is a rigorous methodology for converting these requirements into a final project. The conversion takes place through a series of steps, captured in a form called "house of quality," in which problems such as functionality are related to the characteristics of the product. These steps concern very specific and sometimes sophisticated procedures. Determine the customer's entry objectives here are to understand what your customers want and need from your service/product (critical requirements quality), organize this information, analyze the standards it contains, and develop priorities and strategies. The output is a complete and organized list of customer needs; the highest priority requirements are entered into the project. Identify the customers (external, regulatory, internal) of the product/service in question: those who must meet the needs. You will need to decide if different customer subgroups are likely to have significantly different needs (talk to different "voices"). If so, you'll need to gather information from different segments and look for differences between segments. Typical segmentation factors include economic information (frequency of purchase, revenue generated, etc.), descriptive factors (geographical, demographic, product/service characteristics, industry), and product/service preferences (price, value, characteristics). What you want to do is focus on segments aligned with your company's business strategy. Remember that not all customers represent the same level of value for the company. Perform the customer survey. Use market research, focus groups, interviews, surveys, etc., as appropriate. In addition to proactive customer information, review market research reports, completed assessments, industry reports, competition assessments, access to web pages. Capture your decisions in a customer investigation plan, analysis of information. The goal is to translate voc inputs into customer requirements. The most used tools here are affinity charts (to define attributes) and tree charts (to organize strokes in increasing levels of detail). The client's research plan helps organize team decisions about the market segments you are looking for, how to contact representatives of each segment (interviews, focus groups, surveys) and for what purpose. Qualitative data is an iterative process, which means interpreting and prioritizing. You will likely experience several rounds of data collection while refining your understanding of customer needs. Use QFD to translate vocs into design/performance requirements QFD is a very customer-focused way to design products and services. Emphasize quality "from the outside" (bringing voc to your business rather than relying on internal experts to guess your best. QFD is a more efficient and effective planning method, which reduces costs and downtime) traditional product/service progress stage does not focus heavily on obtaining voc information and planning. This makes the design phase generally to take a long time, as well as the redesign phase, when resolving conflicts. By focusing more, at the planning stage, the QFD process later prevents numerous challenges, creating a short comprehensive cycle and launch into the previous market. Includes a series of analyzes related to building a "quality house" that briefly captures a huge information volume: whatever the customers have said they want, and what importance has been attached to these needs, such as needs they have been converted into valuable want and format which the expected service/product is compared to competition of volatile organic compounds no. Improvement of process and troubleshooting dmaic methodology (identification, measurement, analysis, improvement, and control) related to six sigma is excellent for enhancing awareness about vocs. For example, at the identification stage, the team card creation instructions include obtaining all vocs that are available with genuine information to the defining goal project according to the client's needs. (if the necessary information is not derived, members of the team are required to collect voc data before going through another process).addition, it has become increasingly common for them will be client representatives in their teams. Starting with volatile organic compounds, training and professional skills based on customer needs, in order to really learn how to see a customer, usually will take steps to increase customer awareness among all employees (people involved in lean six sigma projects (possibly internal or external), strategic decision-makers, as well as managers, take nick geisha, vice president of materials and customer service at stanford hospital and clinic (shc) as an example of how this works. There are various logistics departments in the region (mail delivery, etc.) Surrounded by specialized areas such as procurement and contract management. Every day, management affects each client in one way or another. So, our first step was a good understanding of the material and how he helped maximize patient care. "as part of this initiative, which began several years ago, gaze reminded us of a surprise: the patients were in the center, surrounded by a group of carers, and surrounded by the rest of the organization." to achieve the goals of the organization at this points, it means that it is worth what we do last to provide support, "he said. Gayesh tried to achieve this. "our role in procurement is to fully understand the needs of doctors and identify opportunities that doctors cannot determine. Negotiating contracts, technological advances, etc. Our main goal is to use this high level. " as a result, our work is to promote and improve the current development of clinical programs; for example, shcmailroom has more than 100,000 emails per week. There are relatively few specialists (but experts) who are engaged in the following: "if you consider the mail service more high starting point [after all, it's the choice of communications that adds value to your customers] Mr. For these classification methods, pay attention to the number of messages received by this small post office. Please be able to put the wrong answer to the area. Wrong or wrong people or patients. Not everything is reached on time. "therefore, the post office team worked to develop knowledge and improve individual skills to support this feature. The annual assessment allows the material management team to respond to the changing needs of the organization and to analyze or broaden their skills accordingly. This work not only improves the quality of service provided to employees and patients but also enhances the condition of many members of the material management team. For example, for purchasing managers and their employees, the number of daily orders, and the number of products available that allow them to escape the measurement world. It is enough to understand the market and provide doctors with skills and opportunities, and the company's best negotiators will help you provide the best savings for our decisions. A purchase order that can do it more professionally through "this awareness and function to change the profession image amongst representatives of the material group, and most importantly, other shc groups have not been easy." "managing the flow of materials was once a slow idea of the changes that have been made to the institution," says jaish. It is based on a solid foundation of expertise. This is a great feeling for managing materials and staff as it constantly strives to increase the professional competence of colleagues. "

Chapter 7 - CULTURE OF CONTINUOUS IMPROVEMENT

Maintaining the system provided the ideal opportunity to make changes. Maintenance work is generally not mission-critical. However, it is very visible, and the choice was very tactical, as the company contributed directly to setting priorities and is important for short-term business objectives. Maintaining the system was something everyone was concerned about and wanted to work efficiently. And finally, there was a compelling reason to make the change. Everyone was dissatisfied with the existing system. Developers, testers and analysts were exacerbated by the extent of the wasted negotiations and businessmen were very frustrated with the results. We designed a Kanban system that is scheduled to go out every two weeks, scheduled for 1 pm. Schedule a meeting to set priorities with businesspeople every Wednesday and set at ten every Monday. Therefore, the preferred cadence is weekly, and the release cadence is fortnightly. Rhythm selection was determined based on the coordination of transaction costs and activities through cooperative discussions with upstream and downstream partners. Some other changes have been made. We introduced an engineering entry queue with five WIP constraints and added WIP constraints throughout the system's analysis, development, construction and test lifecycles. Acceptance testing, preparation and production readiness remained unlimited. It was not considered limited in terms of capacity and, to some extent, it exceeded immediate political control.

Main effects of the change

The effects of implementing the Kanban system were not surprising at one level but very pronounced at another level. Publications began to be released every two weeks. After about three iterations, they were going smoothly. The quality was good, and there was no need for emergency repair when the new code went into production. Publishing planning and overhead planning expenses have decreased dramatically, and the conflict between the development team and the program management office has almost completely disappeared. Kanban kept his basic promise. High quality, regularly distributed publications with minimal overhead. Transaction and release reconciliation costs have been significantly reduced. The team was working harder and providing their jobs to customers more often. Even more impressive were the side effects. Unexpected effects of implementing Kanban for the development team, we introduced a physical wall with sticky paper on the board in January 2007. We started a morning meeting every day at 9:30 am for 15 days. Physical cards have had a significant psychological impact compared to everything from the electronic tracking tools used by Microsoft. When participating in the stand-up every day, the team members were exposed to some kind of time-lapse photo of the entire workflow. Work items that have been blocked are marked with pink tickets, and the team is more focused on troubleshooting and maintaining the flow. Productivity has improved dramatically. Now that the workflow is displayed on the whiteboard, I started paying attention to how the process works. As a result, we made some changes to the card. My team of managers understood the changes I was making and why, and I made the changes in March. In turn, his team members, individual developers, testers and analysts began to see and understand how things worked. At the beginning of the summer, all team members thought they could propose a change, discussed issues and challenges in the process and observed the spontaneous affiliation of a group of individuals (usually between departments) making the change. If deemed appropriate. Usually, they would have notified the management chain after the fact. Emerging in about six months is the Kaizen culture of the software engineering team. The team members felt empowered. The fear has been removed. They were proud of their professionalism and success and wanted to do better.

Sociological change

In Corbis' experience, there have been other similar reports on the ground. Rob Hathaway of Indigo Blue is the first person to truly replicate these results in the IT group at IPC Media in London. The fact that others were able to replicate the sociological effects of Kanban observed with Corbis is causal and makes me believe that it was not a casual or direct effect of my personal involvement. I thought a lot about what led to these sociological changes. Agile methods have provided transparency for ongoing work over a decade, but teams that follow the Kanban approach seem to achieve a Kaizen culture faster and more effectively than typical Agile software development teams. Often, teams that add Kanban to existing agile practices will see significant improvements in social capital among team members. Why that? My conclusion is that Kanban provides transparency not only for work but also for processing (or workflow). Visualize how work is transferred from one group to another. Kanban allows all interested parties to see the consequences of their actions or non-actions. If the item is locked and someone can unlock it, Kanban will show it. There may be ambiguous requirements. Typically, experts in the field who can clear up ambiguities can expect to receive an email inviting them to a meeting. After the next call, they will hold a meeting, possibly three weeks, according to the schedule. Kanban and visibility topic experts recognize the consequences of omissions and prioritize meetings. Perhaps reorganizing his calendar and schedule this week instead of postponing the meeting for another two weeks. In addition to the visibility of the process, the boundaries of the work performed also cause complex interactions to occur faster and more often. It's not easy to ignore locked objects and just work on something else. The "stop line" aspect of this kanban seems to encourage swarming in the value stream. People with different levels of functions and different tasks will work together to find a solution, support the workflow at the system level and increase productivity, increasing the level of social capital and team trust. Improved collaboration creates a higher level of trust, eliminating fear on the part of the organization. Work in progress boundaries and service classes allow individuals to make their own planning decisions without supervision or supervision. Empowerment raises the level of social capital and shows that managers trust their subordinates in making quality decisions. Managers are exempted from overseeing individual employees and can focus on other things, such as process execution, risk management, team development, and improving customer and employee satisfaction. Kanban significantly increases the level of social capital in a team. Building trust and eliminating fear fosters collaborative innovation and problem-solving. The end effect is the rapid emergence of a kaizen culture.

Viral spread collaboration

Kanban clearly improved the atmosphere of the Corbis software development department, but the results outside his group were most noticeable. It is worth reporting and analyzing how the Kanban viral spreads have improved collaboration within the company.

Negotiation

In the first weeks of the new process, some of the participants await negotiations. They may say, "I know there is only one free space, but there are two small spaces. Can I do both?" This negotiation is rarely allowed. The rest of the board prioritized that everyone played according to the rules. They can answer, "How do they know the little thing? Should I take your word? Or object:" I also have two little things. Select a favourite "Why shouldn't I do this?" I call this "trading period". Because it refers to the negotiation style that took place at the prioritization meeting.

Democracy

Approximately six weeks later, by coincidence, at about the same time that the development team introduced the use of physical advice, the Priority Committee introduced a democratic voting system. They volunteered for this because they were tired of the hassles. Meeting hugs are a waste of time. Improving the voting system required several iterations, but this week it was decided to have a system in which each participant got one vote for each free space in the queue. At the beginning of the meeting, each member proposed a small number of candidates. Over time, handling requests have become more sophisticated. Some people created PowerPoint slides, and others created spreadsheets that described business cases. Later, it was discovered that some members of the organization were inviting colleagues to lunch and lobby. The deal came: "If you voted for your choice this week, will you vote for my choice next week?" Cooperation between business units at the Vice President level, at the heart of the new system of democratic prioritization The level has grown. At the time, we were not aware of this, but the level of social capital throughout the company was growing. When business leaders start working together, people seem to be in the organization. They follow the leader's leadership. Joint actions that combine transparency and transparency create more joint actions. This period is called the period of democracy.

Down in democracy

Everything was very good when it comes to democracy, but after another four months, it seems that democracy could not choose the best candidate. A great effort has been made to introduce e-commerce resources to the Eastern European market. The business case was great, but his candidacy was from the beginning, some doubted the quality of the business case data. After several attempts, this feature was selected and implemented correctly. This is one of the biggest functions managed by the RRT system, and many people have noticed this with involvement. Two months after launch, the Director of Business Intelligence analyzed the revenue data received. This is part of what was promised in the first commercial case, with a payback period of 19 years. Because of the transparency that Kanban provided us, there was a debate about how much valuable bets were invested in that choice when many stakeholders learned about it and were able to make better choices. It was the end of the democratic era.

Collaboration

What replaced him was very remarkable. Remember that the priority committee was mainly composed of company employees and vice-president level directors. They had broad visibility on aspects of the business that many of us were unaware of. Then, at the start of the meeting, they started asking, "Diana, what is the current delivery time?" "We currently have an average of 44 days of production," he said. So they asked a simple question. "What is the most important tactical business initiative in this business in 44 days?" There may be a debate, but generally, there was a quick agreement. "Oh, this is a European marketing campaign launched at a conference in Cannes." What backlogs do you need to support the Cannes event? "Quick search allows you to create a list of six items." There are, therefore, three free places this week. Choose three out of six and contact other users next week. There is little discussion. There were no negotiations or negotiations, and the meeting ended in about 20 minutes, what we called the "collaboration period". This represents the highest level of social capital and trust among the business units achieved during his tenure as senior director of software engineering at Corbis. It is interesting to see how this cultural change emerged and how it had an impact on the whole company, as the employees began to collaborate with colleagues from other business units, under the direction of the Vice President. The change was so serious that the newly appointed general manager Gary Schenk called me at the office and asked if I needed clarification. He said he had seen a new level of collaboration and the spirit of colleagues at higher levels of the company, and that the business units that were previously hostile seem much better. He suggested that the RRT process had something to do with it and asked me if there was an explanation. The system has significantly improved collaboration and, therefore, increased the level of social capital for all parties involved. The cultural side effects of what we now recognize as the capital of K Kanban were completely unexpected and in many ways, counterintuitive. He asked, "Why don't we do all of the major projects this way?" Why? So we decided to implement Kanban in our main project portfolio. It is worth the cost of changing many of the prioritization, planning, reporting and delivery mechanisms that result from Kanban implementations, as Kanban enables Kaizen culture and cultural changes are highly desirable.

CPSIA information can be obtained
at www.ICGtesting.com
Printed in the USA
BVHW081147030822
643612BV00019B/1428

9 781806 034449